Little Science Stories

The Night Sky

By Amanda Gebhardt

The Sun sets. Look up.
There is much to see.

The Moon can rise up high.

It can wax and get big.

It can wane and get small.

The stars shine bright.

They shine deep in space.

Can they see our Sun?

Is it a bright light in their night?

Look! That light shoots
right in the night.

The Sun will rise soon.

It will bring a new day.

Night lights will wait
until it sets again.

Word List

science words

day	stars
Moon	Sun
night	wane
space	wax

sight words

a	the
again	their
our	There
small	to
stars	until
The	

Vowel Teams

/ā/ai, ay, ey	/ē/ee	/ī/igh	/o͞o/ew, oo, ou	/o͝o/oo
day	deep	bright	Moon	Look
They	see	high	new	
they		light	shoots	
wait		lights	soon	
		night	you	
		Night		
		right		

Try It!

Look up at the sky at night.
Draw what you see.

77 Words

The Sun sets. Look up. There is much to see.

The Moon can rise up high.

It can wax and get big.

It can wane and get small.

The stars shine bright.

They shine deep in space.

Can they see our Sun?

Is it a bright light in their night?

Look! That light shoots right in the night.

The Sun will rise soon.

It will bring a new day.

Night lights will wait until it sets again.

Published in the United States of America by Cherry Lake Publishing Group
Ann Arbor, Michigan
www.cherrylakepublishing.com

Photo Credits: © Sergey Nivens/Shutterstock, cover, title page; © Pixel858/Dreamstime.com, 2; © Suyerry/Dreamstime.com, 3; © Calinescu Silviu/Dreamstime.com, 4; © Westlee Appleton/Dreamstime.com, 5; © Triff/Shutterstock, 6; ESA/Hubble & NASA/Judy Schmidt, 7; © Forplayday/Dreamstime.com, 8; © Outer Space/Shutterstock, 9; NASA/Bill Ingalls, 10; © Richard Sheppard/Dreamstime.com, 11; © Derkien/Dreamstime.com, 12; © yaalan/Shutterstock, 13; © Claudio Divizia/Shutterstock, back cover

Cherry Blossom Press is an imprint of Cherry Lake Publishing Group.

Library of Congress Cataloging-in-Publication Data

Names: Gebhardt, Amanda, author.
Title: The night sky / written by Amanda Gebhardt.
Description: Ann Arbor, Michigan : Cherry Blossom Press, [2024] | Series: Little science stories | Audience: Grades K-1 | Summary: "Look up and observe the moon and stars in this decodable science book for beginning readers. A combination of domain-specific sight words and sequenced phonics skills builds confidence in content area reading. Bold, colorful photographs align directly with the text to help readers strengthen comprehension"– Provided by publisher.
Identifiers: LCCN 2023035039 | ISBN 9781668937693 (paperback) | ISBN 9781668940075 (ebook) | ISBN 9781668941423 (pdf)
Subjects: LCSH: Astronomy–Juvenile literature. | Astronomy–Observations–Juvenile literature.
Classification: LCC QB46 .G36 2024 | DDC 520–dc23/eng/20231016
LC record available at https://lccn.loc.gov/2023035039

Printed in the United States of America

Amanda Gebhardt is a curriculum writer and editor and a life-long learner. She lives in Ann Arbor, Michigan, with her husband, two kids, and one playful pup named Cookie.